DeltaScience ContentReaders™

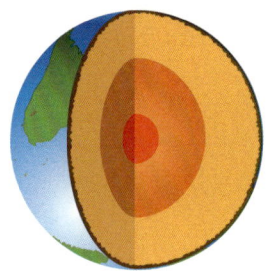

Inside Earth

Contents

Preview the Book 2
What Is Earth Made Of? 3
    Earth's Layers 4
    Earth's Surface 5

How to Read Diagrams 8
How Does Earth Change? 9
    Earth's Plates 10
    How Mountains Form 12
    Studying Earth 14

Cause and Effect 16
What Are Volcanoes and Earthquakes? 17
    Volcanoes 18
    Earthquakes 21

Glossary 24

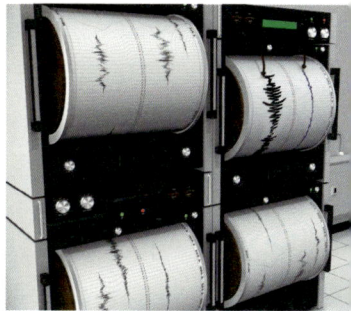

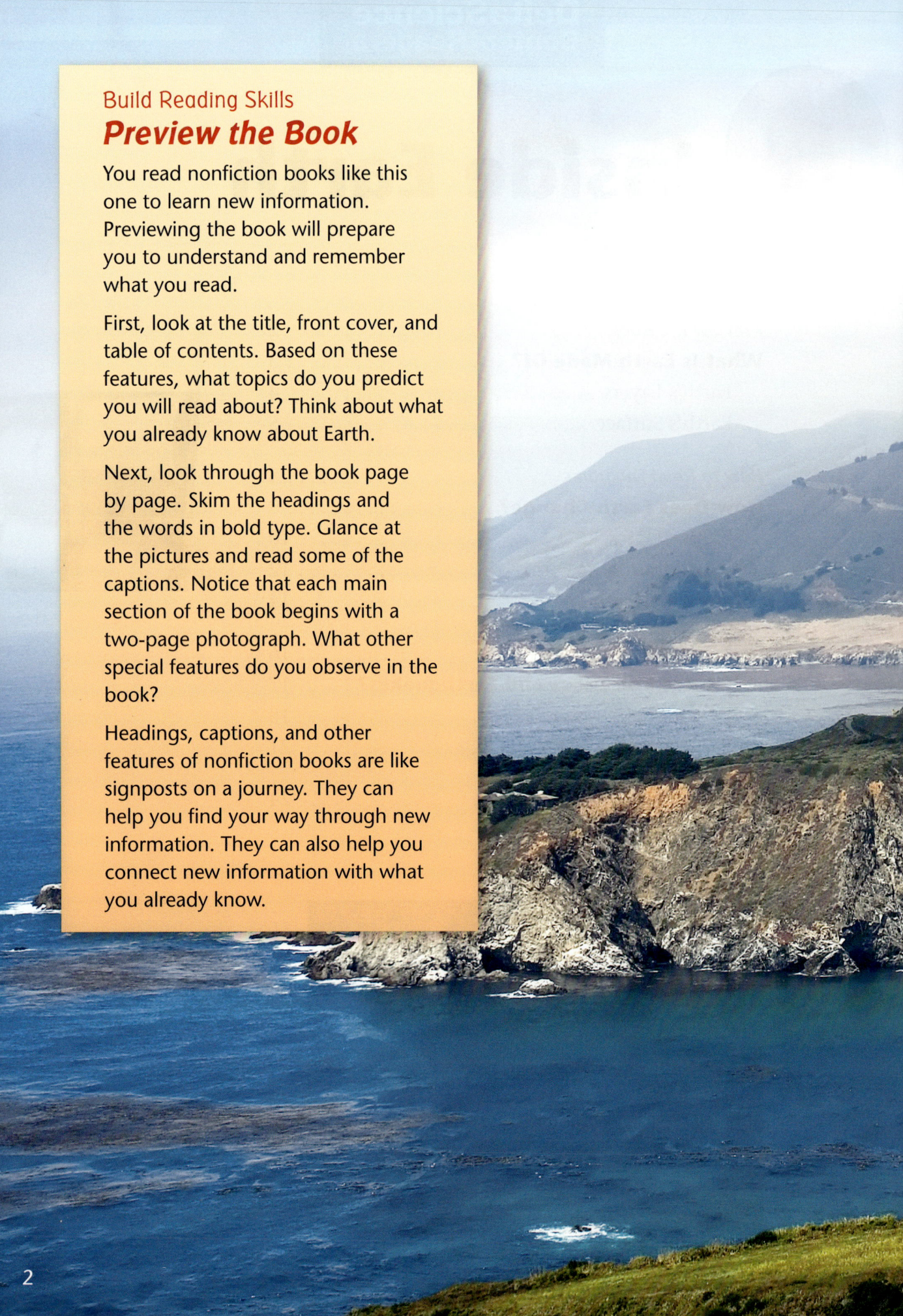

Build Reading Skills
Preview the Book

You read nonfiction books like this one to learn new information. Previewing the book will prepare you to understand and remember what you read.

First, look at the title, front cover, and table of contents. Based on these features, what topics do you predict you will read about? Think about what you already know about Earth.

Next, look through the book page by page. Skim the headings and the words in bold type. Glance at the pictures and read some of the captions. Notice that each main section of the book begins with a two-page photograph. What other special features do you observe in the book?

Headings, captions, and other features of nonfiction books are like signposts on a journey. They can help you find your way through new information. They can also help you connect new information with what you already know.

What Is Earth Made Of?

MAKE A CONNECTION
Earth's land has steep cliffs, rolling hills, tall mountains, and flat plains. Do you think the ocean floor has similar characteristics? Explain your answer.

FIND OUT ABOUT
- Earth's layers
- Earth's surface

VOCABULARY
crust, p. 4
mantle, p. 4
core, p. 5
landform, p. 5
continental shelf, p. 6
continental slope, p. 6
continental rise, p. 6
abyssal plain, p. 6

Earth's Layers

The distance to the center of our planet is more than 6,300 kilometers (almost 4,000 miles). Imagine you invented a machine that allowed you to travel deep inside Earth. What would you discover on your journey? You would probably observe that Earth is composed of rock and metals and that the temperature of the materials inside Earth increases the deeper you go.

Earth has three main layers: the crust, the mantle, and the core. The relatively thin, outermost layer of Earth consists of solid rock. This layer, called the **crust**, covers Earth's entire surface. Almost three-fourths of the crust lies beneath the water of the oceans. A large area of crust that is visible above the surface of the ocean is called a continent.

Scientists who study Earth have identified two main types of crust: continental crust and oceanic crust.

▲ Nearly three-fourths of Earth's crust, or outer layer, is covered by water.

Continental crust, as its name suggests, is the crust that the continents are composed of. Oceanic crust makes up the deep-ocean floor. Continental crust is thicker and lighter, or less dense, than oceanic crust.

Under the crust is another layer of rock, called the **mantle**. Some of the rock in the mantle flows extremely slowly. This occurs because the temperature is high there and the rock is under extreme pressure from the rock above.

Earth has three main layers: the crust, the mantle, and the core. The crust is by far the thinnest layer. ▶

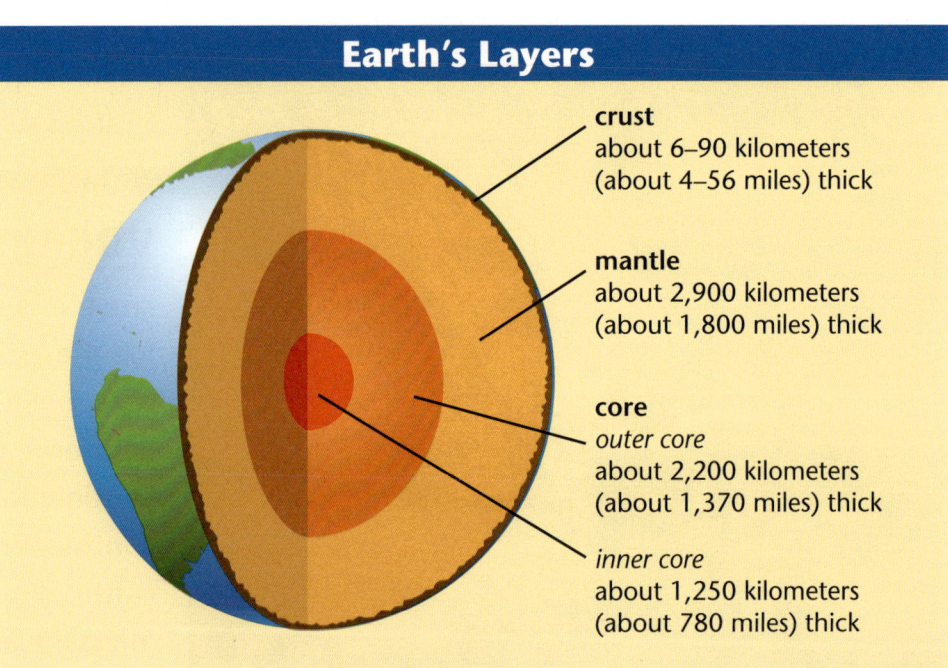

Earth's Layers

crust
about 6–90 kilometers (about 4–56 miles) thick

mantle
about 2,900 kilometers (about 1,800 miles) thick

core
outer core
about 2,200 kilometers (about 1,370 miles) thick

inner core
about 1,250 kilometers (about 780 miles) thick

Beneath the mantle, at the center of Earth, is the **core**. Earth's core is composed mostly of the metals iron and nickel. The core has two main regions: the *outer core*, which is liquid, and the *inner core*, which is solid. The core is the hottest part of Earth. In fact, some studies show that the core could be as hot as, or hotter than, the surface of the Sun.

 Explain how Earth's layers differ from one another.

Earth's Surface

People have long been interested in **landforms**, which are natural features on Earth's surface, and the processes that shape them. Four types of landforms are mountains, valleys, plains, and plateaus. A *mountain* is a landform that is much higher than the surrounding land. A *valley* is a low area of land between mountains or hills. A *plain* is a wide, flat area of land. A *plateau* is flat like a plain but is much higher than the surrounding land.

mountain

valley

plain

plateau

Earth's surface has many natural features called landforms. Mountains, valleys, plains, and plateaus are four types of landforms.

5

Ocean Floor

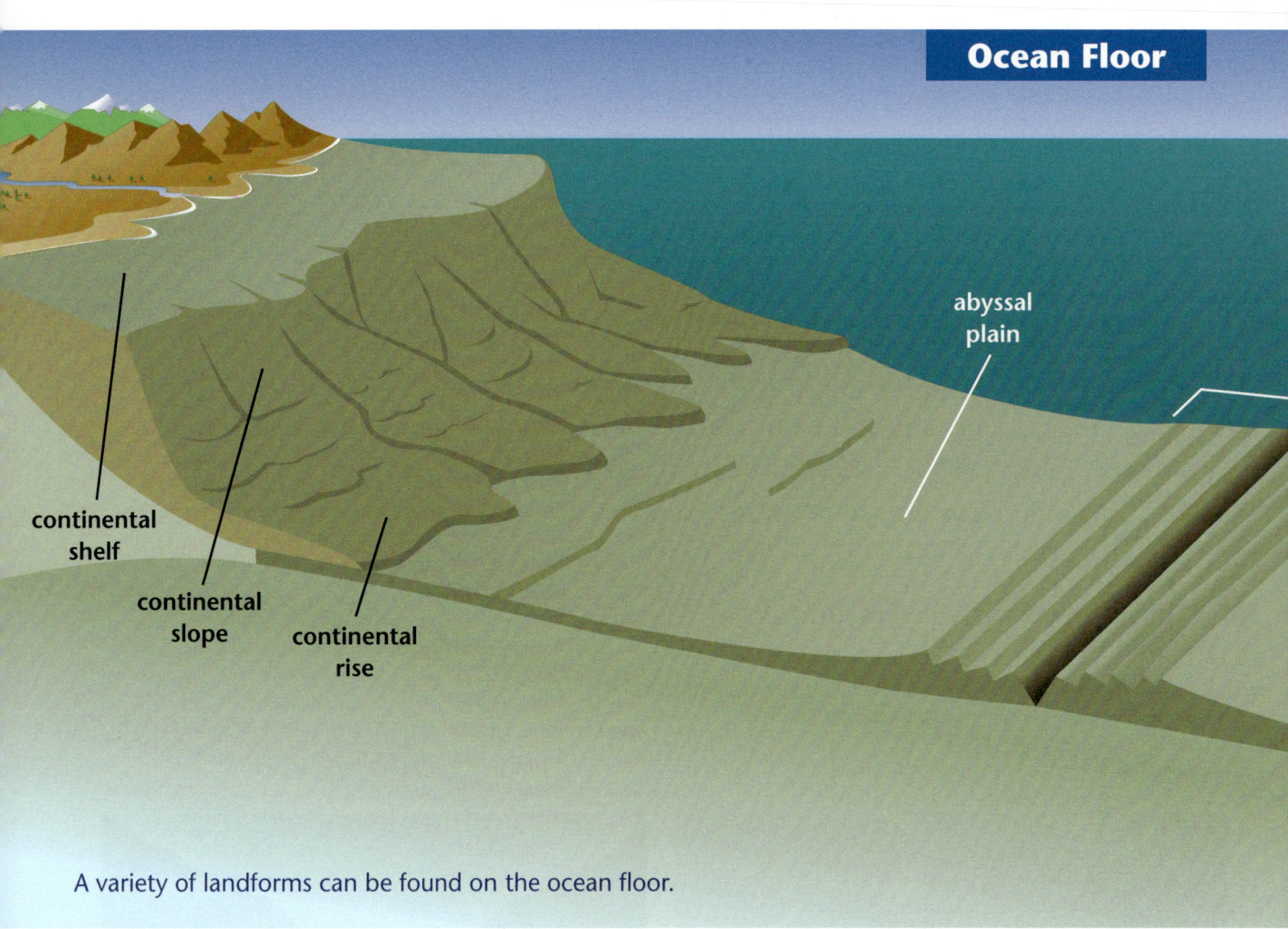

A variety of landforms can be found on the ocean floor.

Much of Earth's rocky surface is beneath the oceans, but even under the water there are many landforms. Gentle slopes, flat plains, tall mountains, and deep valleys all can be found on the ocean floor.

The gently sloping underwater edge of a continent is called the **continental shelf**. Continental shelves vary in size. Some are small and are only about 10 kilometers (about 6 miles) wide. Other continental shelves are large and can be nearly 1,200 kilometers (about 745 miles) wide.

At the edge of a continental shelf, the land drops off sharply, becoming a steep incline called the **continental slope**. Sediment, which includes small bits of rock and other materials, is deposited at the bottom of the continental slope. The buildup of sediment creates a landform known as the **continental rise**. At the continental rise, the land begins to flatten out.

A large portion of the deep-ocean floor consists of very wide, flat areas of land called **abyssal plains**. A thick layer of sediment covers these vast plains.

Scientists have discovered that a chain of mountains called the mid-ocean ridge rises from the ocean floor. Winding around Earth for about 70,000 kilometers (about 43,500 miles), the mid-ocean ridge is Earth's longest mountain chain. Hot, molten rock from Earth's mantle travels up through a

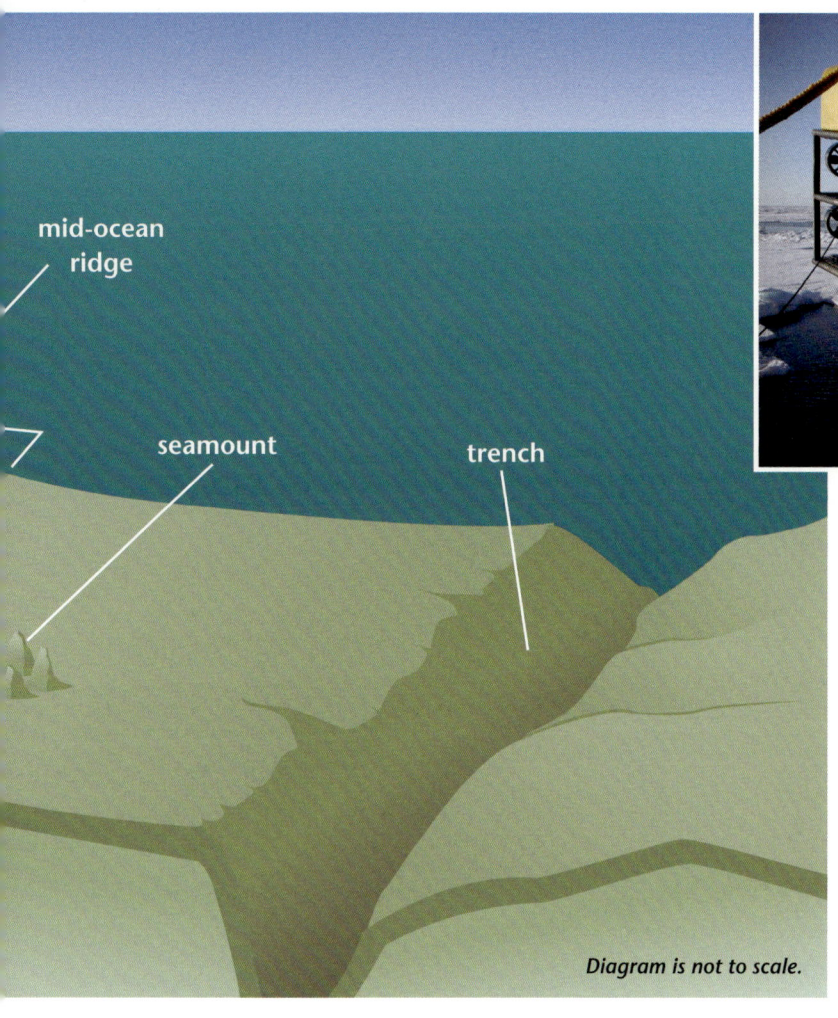

▲ Scientists sometimes use robots to explore the deep ocean.

Diagram is not to scale.

deep center valley, or rift, in the mid-ocean ridge. New oceanic crust is created when this molten rock cools and hardens.

In some locations on the ocean floor, hot, molten rock rising up through the crust has produced underwater volcanoes called seamounts. Some seamounts have flat tops, while others have sharp peaks. A seamount that rises above the surface of the water is known as a volcanic island.

The Hawaiian Islands are examples of volcanic islands.

Some areas of the ocean floor have very deep valleys called trenches. The deepest trench is the Mariana Trench in the Pacific Ocean. The deepest spot in this trench is about 10.9 kilometers (about 6.8 miles) below the surface of the ocean.

 Explain what a landform is, and name four types.

REFLECT ON READING

Before reading, you previewed photographs, diagrams, captions, and other book features. Explain how one photograph or diagram helped you better understand Earth's layers or Earth's surface.

APPLY SCIENCE CONCEPTS

With your classmates, create a poster showing photographs of Earth's landforms. Examples should include mountains, valleys, plains, and plateaus. Label the landforms and tell where on Earth they are located.

7

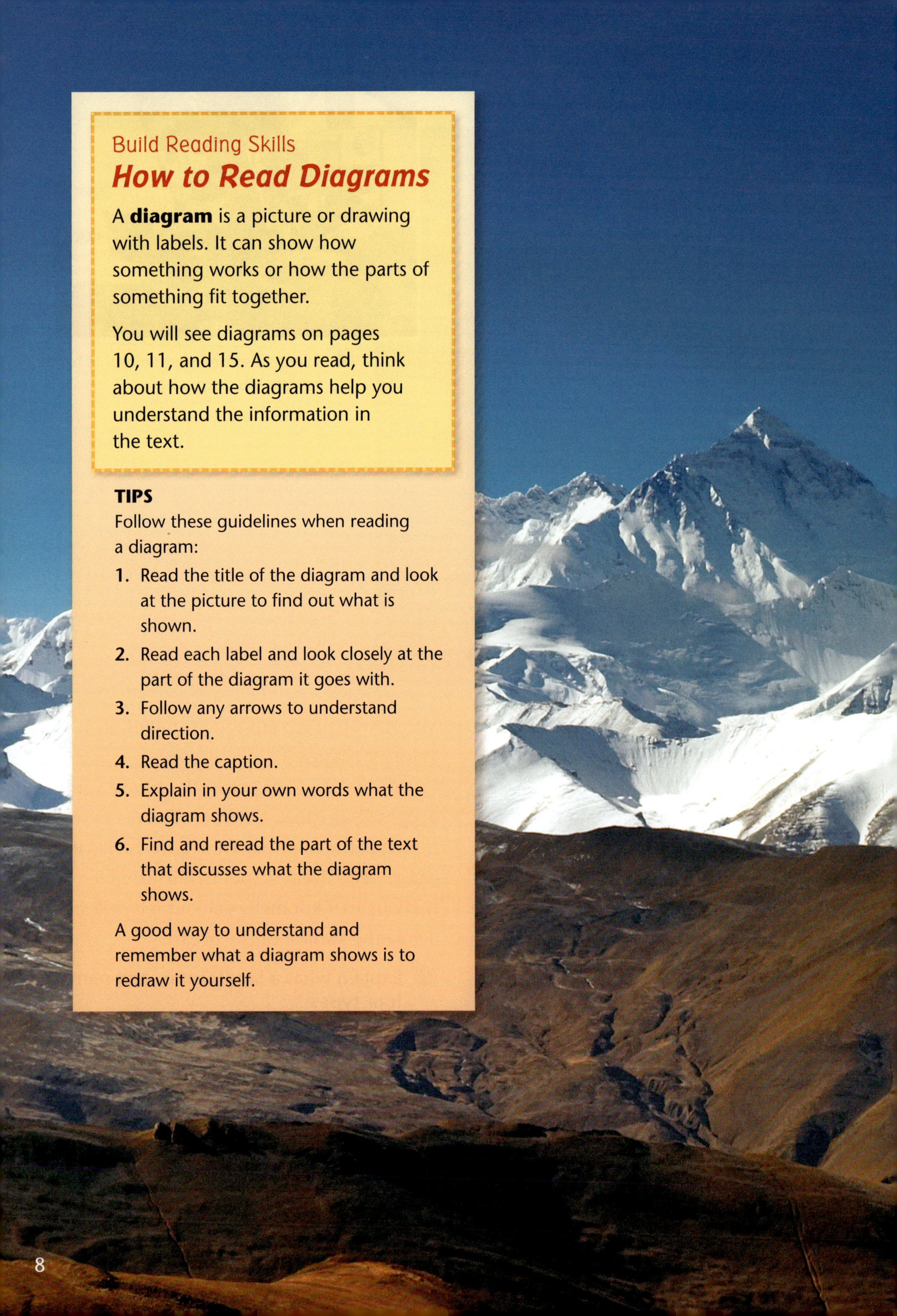

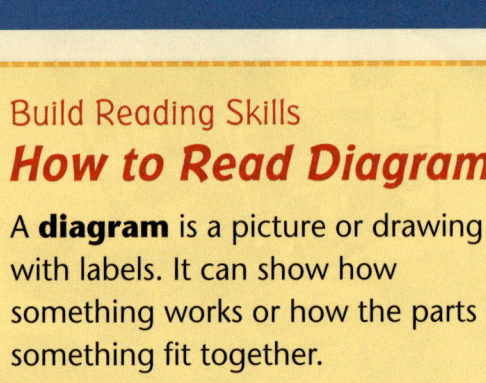

Build Reading Skills
How to Read Diagrams

A **diagram** is a picture or drawing with labels. It can show how something works or how the parts of something fit together.

You will see diagrams on pages 10, 11, and 15. As you read, think about how the diagrams help you understand the information in the text.

TIPS

Follow these guidelines when reading a diagram:

1. Read the title of the diagram and look at the picture to find out what is shown.
2. Read each label and look closely at the part of the diagram it goes with.
3. Follow any arrows to understand direction.
4. Read the caption.
5. Explain in your own words what the diagram shows.
6. Find and reread the part of the text that discusses what the diagram shows.

A good way to understand and remember what a diagram shows is to redraw it yourself.

How Does Earth Change?

MAKE A CONNECTION
The summit of Mount Everest is the highest point on Earth's surface. Mount Everest is in the Himalayas, a group of mountains that stretches across six countries in Asia. How do you think these mountains formed?

FIND OUT ABOUT
- Earth's moving plates and the changes they cause
- ways mountains can form
- ways scientists use rocks and fossils to understand Earth's history

VOCABULARY

plate, p. 10
fault, p. 13
fossil, p. 14

▲ The Appalachians are folded mountains along the eastern edge of the United States and Canada. These mountains formed when plates slowly collided, causing the crust to crumple and fold.

▲ Mount St. Helens in the state of Washington is a volcanic mountain that is part of the Cascade range. It formed when plates slowly collided and one plate sank under the other.

How Mountains Form

Mountains are located on each of Earth's continents. A group of mountains is known as a mountain range. The Adirondacks and Appalachians are mountain ranges found in the eastern part of North America. The Rockies and Cascades are mountain ranges that stretch along the western part of North America.

Mountains form in several different ways. You may have noticed as you read page 11 that plate movements often cause mountains to form. For example, folded mountains form where plates carrying continental crust slowly collide, causing the crust to crumple, fold, and lift up. The Appalachians in North America, the Alps in Europe, and the Himalayas in Asia are examples of folded mountains.

Volcanic mountains can form where plates collide and one plate sinks down under the other. Or they can occur where two plates move apart. In both of these cases, molten rock travels up to Earth's surface, cools, hardens, and builds up over time. The Cascades, which include Mount Rainier and Mount St. Helens in the state of Washington, are volcanic mountains located where one plate is sinking under another. The mid-ocean ridge is a volcanic mountain range located where two plates are moving apart. You will read more about volcanic mountains and how they form on pages 18–20.

Fault-block mountains form where huge blocks of Earth's crust move up or down along a fault. A **fault** is a break in the crust. Faults are often located at or near plate boundaries, but faults also can be found elsewhere on a plate. The Grand Tetons in Wyoming, which are part of the Rocky Mountains, are examples of fault-block mountains.

▲ The Grand Tetons in Wyoming are fault-block mountains in the Rocky Mountain range. The Grand Tetons formed as large pieces of crust moved up or down along a fault.

Dome mountains usually form far away from plate boundaries. A dome mountain is created when molten rock inside Earth pushes up on the crust above it. Over time, the molten rock hardens without ever reaching Earth's surface. This results in a rounded, or domed, landform. The Black Hills in South Dakota and the Adirondacks in the state of New York are examples of dome mountains.

Erosional mountains form when wind, water, or ice erode, or carry away, softer rocks. The harder rocks left behind form mountain peaks. The Catskill Mountains in the state of New York are examples of erosional mountains.

✓ **Describe two ways that mountains can form.**

▲ The Black Hills in South Dakota are dome mountains. They formed when molten rock inside Earth pushed up on the crust, creating a landform with a rounded shape.

The Catskill Mountains in the state of New York are erosional mountains. They are made of harder rock left behind after wind, water, or ice wore away softer rock in the area. ▶

13

Studying Earth

Scientists called geologists study Earth and how it has changed, and keeps changing, over time. Determining the ages of rocks provides geologists with clues about Earth's past.

Some rocks form in layers, with newer rock forming on top of older rock. Knowing this helps geologists determine the age of one rock compared with another. However, movements along plate boundaries and faults can fold, break, lift, and overturn rock. As a result, the newest rock is not always found on top.

Fossils, the remains or traces of living things from long ago, can be used to determine the ages of some rocks. Suppose a rock contains a fossil of a trilobite, a member of a group of animals that lived between 540 million and 245 million years ago. Geologists can conclude that the rock likely formed during that same time period.

Geologists also can perform laboratory tests to determine the age of a rock. Some rocks contain certain materials that change from one form to another over time. Geologists know how much time these changes take. They measure the amount of the original material still left in the rock. Then they can calculate the age of the rock.

The oldest rocks found on Earth are about 4.3 billion years old, and scientists estimate that Earth is about 4.6 billion years old. With the help of fossils and laboratory tests, scientists have developed a timeline of Earth's history called the geologic time scale.

▲ Some rocks form in layers, with newer rock on top of older rock. However, plate movements and movements along faults can fold, break, lift, and overturn rock.

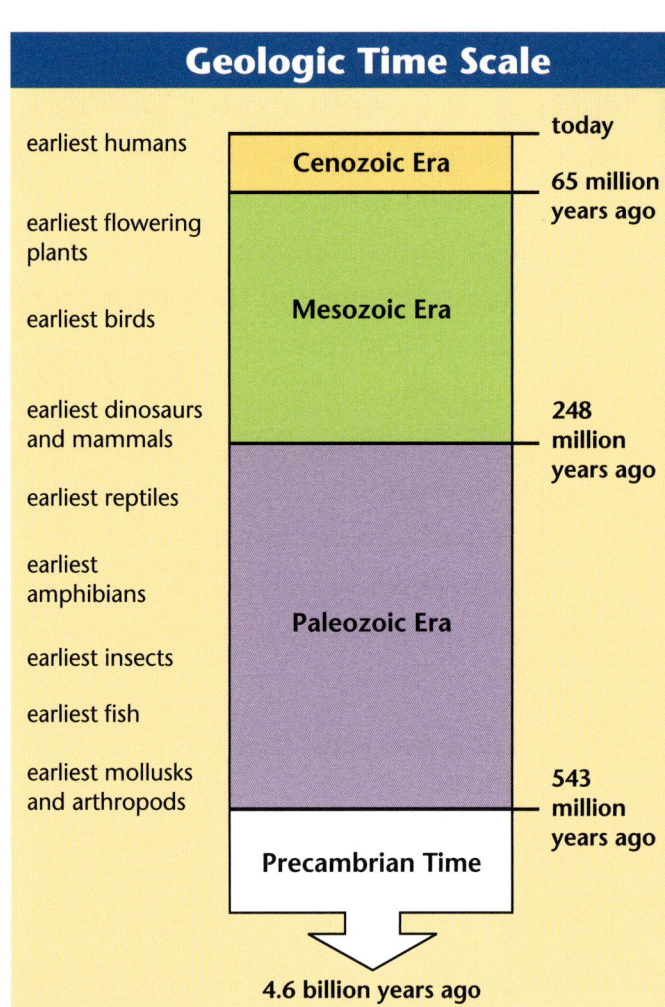

▲ The geologic time scale is a timeline of Earth's history. Precambrian Time is the oldest time period, which began when Earth formed. The Cenozoic Era is the time period we live in.

In 1915, German scientist Alfred Wegener proposed a theory about Earth's continents. He observed that two of the continents—South America and Africa—seem to fit together like puzzle pieces. Wegener also recognized that the types of rocks along the east coast of South America match those along the west coast of Africa. Finally, Wegener observed that the same types of fossils had been discovered in the rocks of both continents. Based on these clues, Wegener theorized that Earth's continents were once joined as one huge supercontinent. Wegener called the supercontinent Pangaea, which is Greek for "all lands."

Scientists now have more evidence about how Earth's continents separated and moved over time and continue to move today. Earth's moving plates include the crust that makes up the continents, and so, as the plates move, the continents move. Plate movements and other Earth processes occurring today also occurred in the past.

 Explain what causes Earth's continents to move.

Earth's Moving Continents

225 million years ago

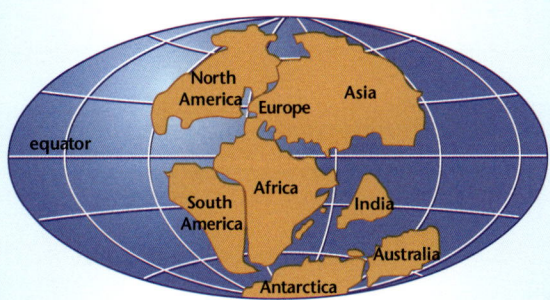
135 million years ago

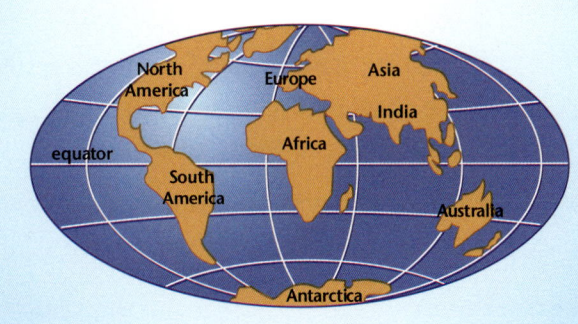
present day

▲ Earth's continents were once joined in a supercontinent called Pangaea, which gradually broke up because of plate movements. These three maps illustrate how the positions of Earth's continents have changed over time.

REFLECT ON READING
Choose one of the plate boundary diagrams from page 11 and redraw it in your science notebook. After your drawing is complete, explain to a partner what it shows.

APPLY SCIENCE CONCEPTS
Select a mountain that you would like to visit. Research the mountain, using books and the Internet. In your science notebook, write the name of the mountain, its location, and an interesting fact about it.

Build Reading Skills
Cause and Effect

A **cause** is the reason something happens. An **effect** is what happens as a result of the cause.

On pages 18–20, you will read about volcanoes. Consider the effects that an erupting volcano can have on Earth's land and air.

TIPS

Recognizing causes and effects can help you understand why events occur the way they do.

- To identify effects, ask, "What happens?"

- To identify causes, ask, "Why does this happen?"

- Words such as *cause, effect, affect, because, why, since, so,* and *as a result* are clues that a cause and effect relationship is being described.

- A cause may have more than one effect, and an effect may have more than one cause.

A cause and effect chart can help you organize your ideas about why things occur as they do.

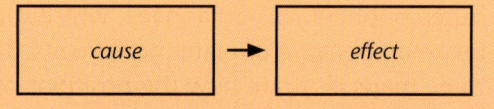

| cause | → | effect |

16

What Are Volcanoes and Earthquakes?

MAKE A CONNECTION
Lava that erupts from a volcano is hot, molten rock that has traveled up to the surface from deep inside Earth. What changes to Earth's surface do you think lava can cause?

FIND OUT ABOUT
- what occurs when a volcano erupts
- three main types of volcanoes
- what causes earthquakes
- measuring earthquakes

VOCABULARY
volcano, p. 18
magma, p. 18
lava, p. 18
earthquake, p. 21

Volcanoes

Recall that molten rock can rise through Earth's crust and form volcanic mountains. An opening in the crust through which molten rock, gases, and other materials erupt, or come out, is called a vent. A vent, as well as the mountain that often forms from materials that erupt through it, is called a **volcano**. Some volcanoes have more than one vent. Hot, molten rock that is below Earth's surface is called **magma**. Once magma reaches Earth's surface, it is called **lava**.

A volcanic eruption can cause rapid changes to Earth's land and atmosphere. Lava, gases, very fine rock particles called ash, and small chunks of hardened lava called cinders can erupt from a volcano's vent. These materials can fly into the air and cover the ground. As a result, living things in the area may be destroyed. Gases, dust, and ash from a large eruption get into Earth's atmosphere. This can affect weather both close to and far away from the erupting volcano.

Volcanic eruptions can be very destructive, but they also build up new land. When lava from an eruption cools and hardens, new rock is produced. As this rock builds up, a mountain can form. Over time, wind and water break down volcanic rocks, and the bits of rock become part of the soil.

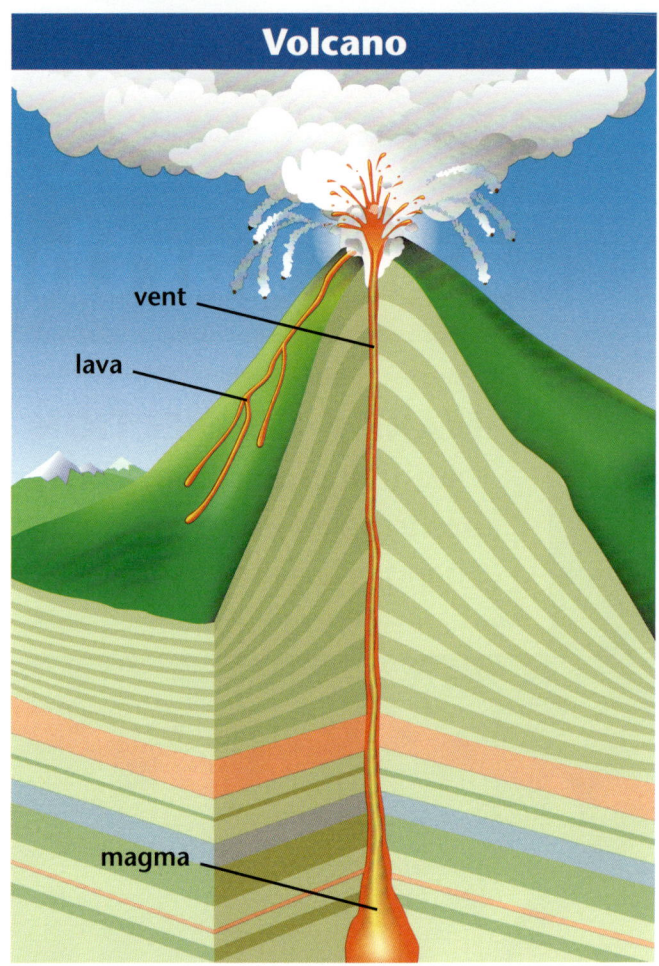

▲ A volcano has one or more vents, or openings, through which hot, molten rock travels up to Earth's surface. Molten rock below the surface is called magma, and molten rock that has reached the surface is called lava.

When lava cools, it hardens into new rock. The black sand on this beach is broken-down bits of volcanic rock. ▶

Volcanoes can be classified by their shape and by the materials they are composed of. Three main types of volcanoes are found on Earth: composite volcanoes, shield volcanoes, and cinder cone volcanoes.

Composite volcanoes are generally tall and wide. They often have very steeply sloped sides that are composed of alternating layers of hardened lava, ash, and cinders. Mount Fuji in Japan, Mount Shasta in California, and Mount St. Helens in the state of Washington are examples of composite volcanoes.

Shield volcanoes also can be tall and occupy a large area, but they are usually not very steep. Instead, they have gently sloping sides that are composed mostly of hardened lava. As the lava cools and hardens, it forms thin sheets of rock that build up around the vent. Kilauea and Mauna Loa on the island of Hawaii are examples of shield volcanoes.

Cinder cone volcanoes are usually relatively small. Most are less than 300 meters (about 1,000 feet) tall. They are not made of lava but form mainly from cinders that erupt from a single vent. Wizard Island in Oregon, Sunset Crater in Arizona, and Paricutín in Mexico are examples of cinder cone volcanoes.

▲ This composite volcano is Mount Fuji, located in Japan. Composite volcanoes have layers of hardened lava that alternate with layers of ash and cinders.

▲ This shield volcano is Kilauea, located in Hawaii. Shield volcanoes are composed mostly of hardened lava and have gently sloping sides.

This cinder cone volcano is Wizard Island, located in Oregon's Crater Lake National Park. Cinder cone volcanoes are composed of cinders, not lava, and they are relatively small compared with composite volcanoes or shield volcanoes. ▶

Each year, about 50 to 70 volcanoes erupt on Earth. Scientists have noticed certain patterns in the locations of the world's volcanoes. For example, many volcanoes are located near plate boundaries. In areas where plates are moving apart, volcanoes form as molten rock travels upward between the separating plates. Other volcanoes form in areas where plates are colliding and one plate is sinking down under the other. Parts of the sinking plate melt, producing magma. That magma often rises and breaks through the crust nearby, creating volcanoes.

The plate boundaries at the edges of the Pacific Ocean are home to a great many of Earth's volcanoes. This region is known as the Ring of Fire.

Not all volcanoes form along plate boundaries—some form above hot spots. A hot spot is an area in the mantle that has a higher temperature than other areas near it. A hot spot melts rock, forming magma that rises and breaks through the crust. When a plate travels over a hot spot, a chain of volcanoes can form. The chain of Hawaiian Islands continues to form as the Pacific Plate slowly travels northwest over a hot spot.

✓ Explain what a volcano is and describe where many volcanoes are located.

▲ More than half the volcanoes on Earth are located at plate boundaries around the Pacific Ocean. This region is called the Ring of Fire.

▲ The island of Hawaii, also called the Big Island, is located over the hot spot that formed the Hawaiian Island chain. The smaller islands to the northwest are older than the Big Island.

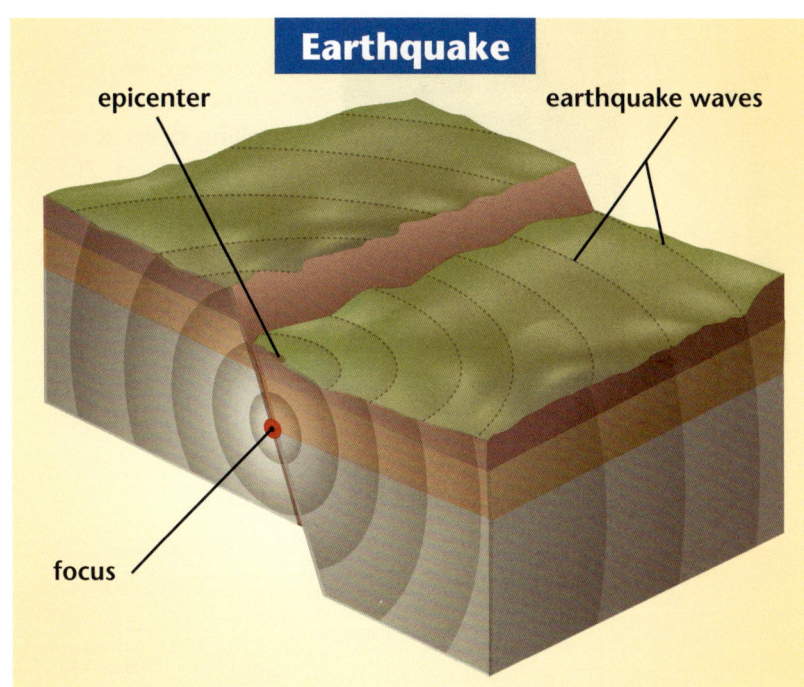

▲ An earthquake occurs when rock underground suddenly shifts, releasing stored energy. The energy travels away from the earthquake's focus, causing the ground to vibrate.

▲ The San Andreas Fault in California is located in an area where plates slide past each other. Many earthquakes occur along this fault.

Earthquakes

Underground rock that is under pressure can suddenly shift. This creates a vibration of the ground known as an **earthquake**. Stored energy is released by the shifting rock, causing the ground to shake. Earthquakes usually occur in the crust near Earth's surface. Most earthquakes are sudden events that last for less than a minute.

Recall that a fault is a break in Earth's crust. Most earthquakes take place at faults near plate boundaries. Many earthquakes in California occur at faults along a boundary where plates slide past each other and pressure builds up. This plate boundary is part of the Ring of Fire.

Earthquakes also can occur where plates collide or move apart.

The point inside Earth where an earthquake begins is called the earthquake's *focus*. Energy released by an earthquake travels out from an earthquake's focus in all directions, moving in repeating patterns called waves. The point on Earth's surface directly above the focus is called the earthquake's *epicenter*.

Scientists have discovered that the energy waves produced by earthquakes travel in different ways through different materials. Studies of how earthquake waves travel have helped scientists determine what materials make up Earth's interior.

▲ Seismographs are instruments that detect, measure, and record the vibrations of the ground during earthquakes.

◂ Some scientists study earthquakes by drilling deep holes in the ground. The scientists record how the rocks in the area behave when an earthquake occurs.

Although thousands of earthquakes occur on Earth every day, most of them are weak and cannot be felt by humans. The amount of energy released by an earthquake is called its magnitude. Generally, larger faults produce earthquakes with greater magnitudes.

Scientists around the world use instruments called *seismographs* to record and study the vibrations from earthquakes. The output from a seismograph, called a seismogram, helps scientists determine the magnitude of an earthquake and its location. A seismogram can also help in finding out the time an earthquake began and measuring how long it lasted.

The magnitude of an earthquake is often described using the Richter scale. The Richter scale ranks earthquakes according to the amount of energy they release. An earthquake with a given number on the scale releases about 32 times more energy than an earthquake with the next lower number. For example, a 2.0 earthquake releases about 32 times more energy than a 1.0 earthquake, and a 5.0 earthquake

Richter Scale

Magnitude	Class
2.9 or less	micro
3.0–3.9	minor
4.0–4.9	light
5.0–5.9	moderate
6.0–6.9	strong
7.0–7.9	major
8.0 or greater	great

▲ The Richter scale is used to describe and compare the magnitudes of earthquakes. The magnitude of an earthquake is the amount of energy released by it.

releases about a million times more energy than a 1.0 earthquake.

Each year, Earth has about 17 major earthquakes (magnitude 7.0–7.9) and about 1 great earthquake (magnitude 8.0 or greater). The strongest earthquake ever recorded occurred in South America, in Chile, in 1960. It measured 9.5 on the Richter scale. The strongest earthquake in the United States occurred at Prince William Sound, Alaska, in 1964. It measured 9.2 on the Richter scale.

▲ Strong earthquakes can greatly damage roads and buildings, change or destroy wildlife habitats, and cause landslides.

▲ The design of this skyscraper in San Francisco, California, helps keep the building stable during an earthquake.

Strong earthquakes can cause a tremendous amount of damage to Earth's surface in a very short period of time. They can cause the ground to crack. Roads, buildings, and property may be destroyed, and power, gas, or water lines may break. As a result of this damage, fires can start. The habitats of plants and animals may be changed or destroyed. In mountain areas, earthquakes can cause landslides. Thousands of tons of rock and soil can slide downhill in seconds.

A strong earthquake on the ocean floor can produce a tsunami, a series of fast-moving waves in the water. When the waves of a tsunami reach shallower water near land, they slow down and pile up, rapidly growing in height. Waves as high as 30 meters (nearly 100 feet) may crash onto the shore, causing great destruction.

Some areas experience a greater number of earthquakes than others. When building in these areas, people can use materials and technology that help keep structures stable. Older buildings and bridges can be reinforced, or made stronger, to reduce earthquake damage.

 Explain what causes an earthquake.

REFLECT ON READING
Create a cause and effect chart like the one on page 16. Write "erupting volcano" in the cause box. Think of effects an erupting volcano can have on Earth's land, air, and living things and add them to the chart.

APPLY SCIENCE CONCEPTS
Do you think that scientists can predict when an earthquake is about to occur? In your science notebook, write a paragraph explaining your ideas.

Glossary

abyssal plain (uh-BIS-uhl PLAYN) a very wide, flat area of the deep-ocean floor **(6)**

continental rise (kon-tuh-NEN-tl RYZE) the underwater section of land that lies at the bottom of the continental slope, where the land starts to flatten out **(6)**

continental shelf (kon-tuh-NEN-tl SHELF) the gently sloping underwater edge of a continent **(6)**

continental slope (kon-tuh-NEN-tl SLOHP) the steeply sloping underwater section of land that lies between the continental shelf and the continental rise **(6)**

core (KOR) the center part of Earth, composed mostly of metals; the outer core is liquid and the inner core is solid **(5)**

crust (KRUHST) the outer layer of Earth, composed of solid rock **(4)**

earthquake (URTH-kwayk) vibrations of the ground that occur when rock beneath Earth's surface suddenly shifts **(21)**

fault (FAWLT) a break in Earth's crust along which movement can occur **(13)**

fossil (FOS-uhl) the remains or traces of a living thing from long ago **(14)**

landform (LAND-form) a natural feature of Earth's surface, such as a mountain, valley, plain, or plateau **(5)**

lava (LAH-vuh) magma that has reached Earth's surface **(18)**

magma (MAG-muh) hot, molten rock that is below Earth's surface **(18)**

mantle (MAN-tl) the layer of Earth between the crust and the core; some of the rock in the mantle flows extremely slowly because of the high temperature and pressure there **(4)**

plate (PLAYT) a thick, moving slab of Earth's crust and part of the upper mantle **(10)**

volcano (vol-KAY-noh) a vent, or opening, in Earth's crust through which lava, ash, and other materials erupt; also the mountain formed from these materials **(18)**